THE BOOK OF EVOLUTIONS

Proof of Evolution, p. 5

Is innovation inevitable? p. 6

Primate Cycle, p. 7

Coherent Brain, p. 9

Grand Evolution, p. 10

Coherent Computer, p. 12

Species Survival, p. 13

Intelligent Species Equations, p. 14

Intelligent Technology, p. 15

Evolution of Technologies, p. 16

Species Extinction, p. 18

Other Equations, p. 19

Neurological Equations, p. 20

Social Models, p. 21

Linear Models, p. 22

Coherent Models, p. 28

Pragmatic T.O.E. p. 32

Posthumanism, p. 33

BIO, p. 40

Devoted to the Millennial Darwinists...

Who make this stuff look fresh.

© As a compilation 2022. Some materials previously made available under copyright control 2013, 2014, 2015, 2016, 2017, 2018, 2019, 2020, 2021, 2022.

THE BOOK OF EVOLUTIONS

By Nathan Coppedge

HISTORY OF IDEAS PAPER
START ANYWHERE, ARRANGE CHRONOLOGICALLY
These refer to rough dates of each invention as a science.

Technological Complex is	Technological Complex is
Technological Simple is	Technological Simple is
Artistic Simple is	Artistic Simple is
Artistic Complex is	Artistic Complex is
Cosmological Complex is	Cosmological Complex is
Cosmological Simple is	Cosmological Simple is
Physical Simple is	Physical Simple is
Physical Complex is	Physical Complex is
A New Concept is	A New Concept is
Technological Complex is	Technological Complex is
Technological Simple is	Technological Simple is
Artistic Simple is	Artistic Simple is
Artistic Complex is	Artistic Complex is
Cosmological Complex is	Cosmological Complex is
Cosmological Simple is	Cosmological Simple is
Physical Simple is	Physical Simple is
Physical Complex is	Physical Complex is
A New Concept is	A New Concept is

PROOF: (1)No Nc --> Limited complexity (brain science), Limited complexity - -> No Tc (technology),
No Nc - -> No Tc (hypothetical syllogism), Nc --> Tc (negation or double-negation), Nc, Tc (2)Tc - -> Ts(Ockham)
else No Tc., Tc, therefore Ts (3)All Ts (includes As), Sufficient Ts therefore sufficient As(4) Ts --> As,
(c, s) measure same thing., Tc --> Ac, Tc, Ac(5) Ac is a symbol for Cc, A symbol is a description.,
Ac --> Description Cc (Substitution)., Description Cc equivalent to Cc (Descriptive materialism), Ac, Cc
(6)Ac --> Description Cc, (c,s) measure same thing., As --> Description Cs, As, Cs (Descriptive materialism).
(7) Cs = Ps Existential Tautology., Ps (8) Cc, Cs, Ps, (c, s) measure same thing., Pc (combination)
(9)No Nc - -> No Tc (from 1), Tc (from 1) supported by Pc (from 8), Nc (modus tollens and negation applied twice).

HOW TO MAKE INNOVATION INEVITABLE?

Once you have more than about five geniuses, that's one way.

Another way, is if you have several research corporations competing with one another.

Another way is if you have artists and money.

Another way is if you have a lot of logical thinking like systems and visualization and ask for suggestions.

Another way is having really smart genes and helpful education.

Another way is to provide monetary incentives for military research, or other survival tasks like building bridges or building safer buildings.

Another way is if you have very long-lived people who exhibit youthful properties at an old age.

Another way is if you have a dynamic product like computers, writing, or the Enchiridion to motivate people.

Another way is if people are addicted to life, like in the 1920's or during the Greek Golden Age.

Another way is if there is a respected institute that amuses everyone, like sometimes happens in London.

Another way is if people band together using some sort of adaptation, like rope-climbing or pit-traps.

Another way is by necessity if you're part of a successful species.

PRIMATE CYCLE: In the following dialectic: be strong, or genius, or help civilization survive.

DIALECTICS OF PRIMATES:

SITUATION 1

Someone is strong enough to survive.

SITUATION 2

Someone is dominant enough to survive better.

SITUATION 3

Smart people become 'sensitive' because they are selecting for 'smart'.

SITUATION 4

Reproduction becomes stupid or easy.

SITUATION 5

With reproduction triggering sensitivity, appearances of stupidity dominate.

SITUATION 6

Things are taken-for-granted, and possible cognition arrives.

SITUATION 7

People fail enlightenment.

SITUATION 8

If things aren't okay, sanity is tested very carefully or people go crazy.

SITUATION 9

Either there isn't much of a problem to begin with or people are in massive social denial.

SITUATION 10

The voice of the dominant characteristic becomes important

SITUATION 11

Political conservativism begins to wear thin.

SITUATION 12

There is a period of indulgence.

SITUATION 13

Pleasure's heroes seek protection.

SITUATION 14

If there are geniuses born they may appear here.

SITUATION 15

Sensitive people appear under the wing of the geniuses.

SITUATION 16

Every time the poets are massacred it takes time for civilization to recover.

SITUATION 17

If civilization does not survive, then helpful people ultimately die.

—**What is the most likely endpoint of human evolution?** (...)

THE COHERENT BRAIN ED 1

Re-assessment of human ideas:

1. Physical luck. 14. Regular luck.

2. Greed for works. 5. Greed for ideas. 7. Regular greed.

3. Sufficient ideas. 6. Idea. 10. Obviated idea.

4. And 9. Mental sensations.

5. (Skip).

6. (Skip).

7. (Skip).

8. Mental-physical works.

9. (Skip).

10. (Skip).

11. And 12 Physical art.

12. (Skip).

13. Madness.

14. (Skip).

GRAND EVOLUTION CONCEPT: 20X DIRECT EVOLUTION, HOW TO EVOLVE: Simply find a better overall concept based on the immediately previous concepts: Quite possibly we need a higher-category Theory of Everything:

OUTER SPACE

- It is cold or hot.
- It contains black holes.

BLACK HOLES

- They are big and cold.
- They can destroy stars.

COLD TEMPERATURES

- Freezes everything solid, so it breaks down into powder.
- Prevents most chemical reactions.

BURNING STARS

- They run out of fuel and might turn into a black hole.
- They burn very far away.

HIGH TEMPERATURES

- Painful and radioactive.
- Often persists where conditions apply.

PLANETS

- Hard to find, exist in outer space.
- There are problems on planets.

TOXIC GAS

- Easily created, and hard to destroy.
- Hurts biological life.

MOLTEN LAVA

- Unpredictable, destructive.
- Burning things.

BURNING ASH

- Clouds of ash can cause semi-permanent cold weather in rare cases.
- Ash covers the ground, making transportation difficult.

POOL OF ACID

- Melts through protective coverings.

- Painful to animals.

INDUSTRIAL WASTE

- Can involve dangerous chemicals or radiation.
- Can hurt biological life.

MOUNTAINS

- Big and slightly dangerous
- Requires energy to climb.

POND WATER:

- If you don't know how to swim.
- Also, too many dangers / parasites.

FORESTS

- Sometimes poisonous plants or dangerous animals.
- Easy to get lost / hard to survive.

GRASSLANDS

- Hard to find food and shelter.
- Hard to hide from animals and humans.

DESERTS:

- Too dry and hot or cold at night.
- Dangerous animals and environments.

VIRUSES AND DNA

- Microscopic and infiltrate the body.
- May require special medicine.

FOOD

- If you don't have it you starve or get a headache.
- Sometimes poisonous.

ANIMALS AND MONSTERS

- Defensive and offensive.
- May carry pestilence.

HUMANS AND ROBOTS

- Sometimes attack / carry weapons.
- Can be tricky / unhelpful. —<u>Wider Morality</u>. [The wider morality is based on the earlier <u>Pond Logic</u>]

COHERENT COMPUTER

UNIVERSE DNA:

BCAD or DCAB AND (CDBA or ADBC)

AND DACB or BACD AND (ABDC or CBDA)

OR

BCAD or DCAB AND (ABDC or CBDA)

AND DACB or BACD AND (CDBA or ADBC)

Also translates as:

GIVEN QUESTION (A) is C: BD THEN BCAD and / or DCAB

GIVEN QUESTION (B) is D: CA THEN CDBA and / or ADBC

GIVEN QUESTION (C) is A: DB THEN DACB and / or BACD

GIVEN QUESTION (D) is B: AC THEN ABDC and / or CBDA

SPECIES SURVIVAL EQUATION

NEW SPECIES SURVIVAL EQUATION: number of dominant species that survive equals 5, 9, 17, 34, 68 with 25 categories of math and 2 conceptual dimensions, 20 modular ideas and 11 dimensional TOE. The same equation often with different resulting numbers governs species under alternate TOEs. the equation is mostly a function of the conceptual dimensions and the number of categories of math.

FORMULAS

D = Spatial dimensions, also called conceptual dimensions.

TC = Total Categories (Mathematical number theory categories, like zero, finite or infinite)

PRIMARY SPECIES EQUATION = (TC-((POWER(TC,(1/D))-D)^D)+1)

SPECIES EQUATION - 2 = (((TC-((POWER(TC,(1/D))-D)^D)/4)+1)

SPECIES EQUATION - 1 = (((TC-((POWER(TC,(1/D))-D)^D)/2)+1)

SPECIES EQUATION +1 = ((TC-((POWER(TC,(1/D))-D)^D)+1)*2)

SPECIES EQUATION + 2 = ((TC-((POWER(TC,(1/D))-D)^D)+1)*4)

This is number of very distinct alien species or alternately dominant species in the observable universe by various objective standards.

2nd Dimension (25 TOE Categories): 5, 9, 17, 34, 68

3rd Dimension (8 TOE Categories): 3.25, 5.5, 10, 20, 40

3rd Dimension 64 TOE Categories: 16.75, 32.5, 64, 128, 256

4th Dimension, 256 TOE categories: 65, 129, 257, 514, 1028

8th Dimension, 256 TOE categories: -419839, -839679, -1679359, -3358718, -6717436 (minuses)

8th Dimension, 2097152 TOE categories: (524257.3959, 1048513.792, 2097026.583, 4194053.167, 8388106.334)

INTELLIGENT SPECIES EQUATIONS

Basic Systems equation:

Given evolution, new efficient species Requires new basic systems.

Proof: If an old species requires new systems, it has not completely evolved. New species must do something fundamentally different, and efficient species must use systems.

[Note: This could be applied recursively to find differences across planets].

Collective Species equation:

If no new basic systems, No new efficient species, or existing species must evolve.

Proof: Species that evolve are efficient (efficiency uses systems), and new species must be fundamentally different

INTELLIGENT TECHNOLOGY PAPER
SELECT:

INVESTIGATE: FIRST PICK THE OPP CATEGORY OF YOU
THEN EARLY LIST IF NOTHING TRIVIAL
LATER LIST IF EVERYTHING TRIVIAL

1. Genius

Exponentially Efficient Genius ---> Compound Efficiencies,
A Little Genius ---> Ideas,
Something Very Intelligent --> Strategy,
The Greatest Idea --> Ambition,
A New-Everything-Genius ---> Do everything.

2. Matter

Exponentially Efficient Matter ---> Physical Calculation,
A Little Material Genius ---> Chemicals,
Something Materially Intelligent ---> Adaptation,
A Great Matter-Idea ---> New Forms of Matter,
A New-Matter-of-Everything ---> Unifying Physics

3. Organization

Exponential-Efficient Organization ---> Hyper-Organized,
A Little Physical-Organizational Genius ---> Classification System,
Something Physically Intelligent and Organized ---> Functionality,
A Great Physics Organization Idea ---> Physical Classification,
A New Organization-of-Everything ---> A Schematic

4. Solutions

Exponentially Efficient Solutions ---> Automatic Calculation,
A Little Physical Genius Solution ---> Knacks,
Something Physically-Intelligent Solution ---> Mechanism,
A Great Physical Solution Idea ---> General Physical Solutions,
A New Solution-to-Everything ---> A Practical Invention

Reproducible under Nathan Larkin Coppedge

EVOLUTION OF TECHNOLOGIES

Theory of Anything Tech: (Eff - Diff) / (Eff + Diff)

- Eff = 3 | Diff |
- Primary Values: Eff 3 + Diff 1 = Results 2.5 to 4
- Alternate Values: Eff 3 + Diff -1 = Results 2

Antitheory Tech: (Diff + Eff) / (Diff - Eff)

- Diff = 3 | Eff |
- Primary Values: Diff 3 - Eff 1 = Antitheory 0.5 to 2
- Alternate Values: Diff 3 - Eff - 1 = Antitheory 4

Efficiency and Anti-Efficiency Tech: (Results + Diff) / (Results - Diff) AND (-Results - Diff) / (-Results + Diff)

- Results = 2 | Diff |
- Primary Efficiency Values: Results 2 - Diff -1 = Efficiency 2 to 3,
- Alternate Values: Results 2 - Diff 1 = Efficiency 1
- Primary Anti-Efficiency Values: Results -2 + Diff 1 = Anti-Efficiency 0 to -1
- Alternate Values: Results -2 + Diff - 1 = Anti-Efficiency -3

Difference and Anti-Difference Tech: (Results + Eff) / (Results - Eff) AND (-Results - Eff) / (-Results + Eff)

- Results = 2 | Eff |
- Primary Difference Values: Results 2 - Eff -1 = Difference 2 to 3,
- Alternate Values: Results 2 - Eff 1 = Diff 1
- Primary Anti-Difference Values: Results -2 + Eff 1 = Anti-Difference 0 to -1
- Alternate Values: Results -2 + Eff -1 = Anti-Difference -3

Dimensional and Anti-Dimensional Tech: (#Antiforces - #Forces) / (#Antiforces + #Forces) AND (#Antiforces + #Forces) / (#Antiforces - #Forces)

- +/- Forces = 0.333 | Antiforces |
- Primary Dimensional Values: Antiforces 2 + Forces 1 = Dimensions 2 to 3
- Alternate Values: Antiforces 2 + Forces -1 = Dimensions 1
- +/- Antiforces = 0.333 | Forces |
- Primary Anti-Dimensional Values: Antiforces 3 - Forces 1 = Anti-Dimensions 0.5 to 2
- Alternate Values: Antiforces 3 - Forces -1 = Anti-Dimensions 4

Forces and Negative Forces Tech: (#Dimensions + #Antiforces) / (#Dimensions - # Antiforces) AND (-#Dimensions - #Antiforces) / (-#Dimensions + #Antiforces)

- Optimized tentatively at Special Efficiency = 2 (positive antiforces), 0.5 (negative antiforces), +/- Antiforces = 0.333 | Dimensions |
- Primary Force Values: Dimensions 3 - Antiforces - 1 = Forces 2.5 to 4
- Alternate Values: Dimensions 3 - Antiforces 1 = Forces 2
- Primary Negative Force Values: Dimensions -3 + Antiforces 1 = Neg Forces -0.5 to -2
- Alternate Values: Dimensions -3 + Antiforces -1 = Neg Forces -4

Antiforces and Negative Antiforces Tech: (#Dimensions + #Forces) / (#Dimensions - #Forces) AND (-#Dimensions - #Forces) / (-#Dimensions + #Forces)

- Optimized tentatively at Special Efficiency = 2 (positive forces), 0.5 (negative forces), +/- Forces = 0.333 | Dimensions |
- Primary Antiforce Values: Dimensions -3 + Forces -1 = Antiforces -2.5 to -4
- Alternate Values: Dimensions -3 + Forces 1 = Antiforces -2
- Primary Negative Antiforce Values: Dimensions -3 + Forces 1 = Neg Antiforces -0.5 to -2
- Alternate Values: Dimensions -3 + Forces -1 = Neg Antiforces -4

SPECIES EXTINCTION EQUATION:

You have a paradigm, then you do something just for demonstration.

For example:

WITH ANCIENT BEASTS:

- You just want to be big XXX dinosaurs.
- You just want to swim XXX some fish species.

WITH MAMMALS:

- You just want to be warm XXX woolly mammoths.
- You just want to hunt XXX snow leopards, etc.
- You just want to be unique XXX birds of paradise.

WITH HUMANS:

- You just try to survive XXX Neanderthals.
- You're half beast XXX beast-men.
- You just want to survive in the wild XXX feral children.
- You just want to be a robot XXX cyborgs. <u>Details on why someone might not want to be a cyborg or gene-enhanced etc.</u>

Casualties Watchlist:

- Quantum computers can be beat by conventional computers or a new invention ('Linnaean stem problem'). XXX quantum computers or expect technology to slow down.
- Education requires stimulation, fairness, and completeness. Coherence is the simplest, most fair-minded way to stimulate students. XXX incoherent approach to education or expect stupider students.
- Drugs benefit from coherent approaches. XXX serotonin cycle approach or expect harder drugs and more depression.
- Students are learning about mathematics. XXX non-dimensional (non-coherently-organized) approaches or expect a reversal against mathematics.
- Too much obsession with outer space?
- It could be pernicious for A.I. that humans may have developed morality after developing evil.

OTHER EQUATIONS

"Equity": Given resources, if poverty isn't the solution, then certain populations probably have a right to reproduce, or key inventions haven't been made.

Zeeman Effect: Zeeman effect in biology: The second things to appear, if they appear for a reason, occur as a result of the earlier things (such as organs). Thus, centralized things are a result of a central process, and secondary things are a result of a secondary process, or the Zeeman effect is broken. These theories apply from the simplest things to the most complex. So the big lemma is simply nature itself.

MORTALITY: $-2(D) = $ X'd Entity

NEUROLOGICAL EQUATIONS

Put to an incomplete arbitrarily complicated task, however, the brain can be very limited.

Either life is peaches, or you're doomed, or you don't understand psychology.

If you're wrong, then you don't understand psychology.

1000 IQ Formula= Stimulating yourself to be Einstein ^ 14.

Bad theory is worse preferences.

Synapse formation: New synapses are formed from the obvious deformation of the disintegral.

Nature of Neurons: The question of the many kinds of neurons is answered by the singular non-typed genera of the body.

Neural Code: It requires a code.

Neuronal Free Will: The freedom of the sensory system is determined by reactions against systems.

Evolution of the Brain: The question of the brain is answered by 'why not be immature?' The quality of individual brain substances are determined by an indeterminate number of immature groups.

Organization of the Brain: The senses are chemical reactions disintegrating, the rational purpose was imposed through organization, which is a kind of extraneous characteristic, an offshoot of chemical reactions disintegrating.

The brain location of costs-to-benefits is embodied with qualities of double-negative benefits, this has the result of creating brain plasticity.

Dreaming: Curiosity about thinking and curiosity about visual vision.

SOCIAL MODELS

So, essentially, you want a good system with ethical reactionaries, authentic crazies, ethical actors, democratic experts, conservatives, or something less expensive. —writing on Charles Darwin

Coherent Motivation Model: I've done everything that I don't remember. Could mean I've done everything I am not still finishing. Because if I remember it, it's something that happened in some way.

Human Model: From a human standpoint, the universe can be summarized is a mixture of virtual reality, emotions, urbanization, and limits.

Urban Model: That selfish men are located inside a city, and altruistic women are located on the outskirts of a city. —Based on: What would be an example of a saving people complex? And is that really a good thing?

Gendered Thinking Model: Women are frequently interested in specialism like the human body, medicine, biology, neuroscience, and psychology, men are frequently interested in generalism like philosophy, logic, systemization, history, music, anthropology, and politics.

Dishonesty Model: By analogy, something more complex simply evokes more complex responses of the same kind: a *literal-complex* response. If humans are dishonest, that is still a *dishonest*, literal response. —James-Lange Theory (literal response to television)

Drug Model: Borderline: May for example, be someone who thinks they love coffee but are really fairly moderate and have never been exposed to the hardest drugs. Depression: May for example, be someone who drinks a lot of coffee who was exposed to hard drugs much later in life. Schizophrenia: May for example be someone who hates coffee but was exposed to hard drugs second-hand. Psychotic: May for example, be someone who was exposed to hard drugs early, then tried coffee and quit, or who was exposed to a large amounts of second-hand drugs intermittently. A normal person is someone who was never exposed to coffee and never exposed to drugs, or someone moderate with a high tolerance for drugs. Other categories: good medication (oh so rare). —Mental Health Toxicity Model

LINEAR MODELS

3-POINT CULTURES (EQUIVALENT CULTURES)

DINOSAUR CULTURE

- Life is short.
- One thing is good.
- One thing is bad.

EARLY MEN

- You get taught by your mother.
- You take a journey just for you.
- You learn something at some point.

SUMERIAN CULTURE

- There is one big event that doesn't take long.
- You understand it, and you appreciate it like it's nothing.
- You feel more unique as a result.

GERMAN CULTURE

- What you do matters to you.
- It's the absolute, ultimate.
- You are summarily rejected.

CHINESE CULTURE

- It is for very original people.
- You may need a Babel Fish or it is less likely you are original.
- It can be corrupting.

EGYPTIAN CULTURE

- There is something magical about it.
- I always seemed to learn something.
- It has a dark side that could cause trouble.

GREEK CULTURE

- They will use you.
- They think nothing of your gods.
- What happens is always very important.

FRENCH CULTURE

- You must become a politician.
- Life does not end well.
- It is all about people.

ENGLISH CULTURE

- Takes time to get to know it.
- I always seem to have status and privilege.
- I found it was easier to collect numerous intellectual ideas when I was English.

Note: YY says English culture changes over time.

AMERICAN CULTURE

- You can think as well as the English.
- It is a boring adventure.
- Life is usually painful, but the food is good.

3-Point Cultures.

2-POINT CULTURES (DOMINANT CULTURES)

Chinese :

- Martialing the arts.
- Overwhelming forces.

Phoenicians :

- Overwhelming forces.
- Religious traditions.

Egyptians :

- Religious traditions.
- Appealing culture.

Greeks :

- Appealing culture.
- Enslaving others' cultures.

Romans :

- Enslaving others' cultures.
- Consolidating power.

French:

- Consolidating power.
- Robbing the people.

British :

- Robbing the people.
- Industrialization.

Americans :

- Industrialization.
- Mass production.

Rich culture:

- Mass production
- Perpetual motion?

...

LINEAR INTELLECTUAL DEVELOPMENT
CONTENT PROGRESSION

1. Take a good example, and create several versions of it.

Example: *My primary treatment is to translate or expand Ockham to include philosophical razors designed to provide standards for how to do philosophy.*

2. With your experience, extract a higher principle from the original example.

Example: *One perspective perhaps related to this is that Ockham can also have a 'higher translation' in terms of logical or mechanical (etc.) efficiency.*

3. Greatly improve the higher principle by adding another factor, for example, simply doubling it.

Example: *We can then use the principle of efficiency ingeniously to arrive at exponential efficiency.*

4. Now, use the improved higher standard as a platform for a body of very new ideas.

Example: *Exponential efficiency can then be used as a platform concept for masterful fulfillment of the logical and mechanical criteria.*

5. Find the best general categories within the new system / platform.

Example: *This leads to the general concepts of preferred knowledge and continuous motion machines,*

6. Now translate the general categories using your understanding of the general and specific meaning.

Example: *...equals objective knowledge and perpetual motion.*

LINEAR VERSION OF A COHERENT MODEL

If we premise that the TOE reduces to 5/32 (though this appears to be asymmetric, associated with Universe 7) it appears we can select 'essential cycles' within columns of any row, creating a rubric schematic in which the first column containing the TOE and perpetual motion machines is given functionalistic preference over the other columns... If we choose we may iterate a given column in a highly repetitive fashion in order to arrive at new types of ideas which follow the pattern of corresponding with the categories of the first column taken downwards repeatedly without retracing. This process need only be followed several times from a given 'Essence' (such as 5/32) to realize that the cycle is always noticeably the same, though it may vary in quality and energy from column to column, and like I said the leftmost column in my usual order which has the highest mathematical RESULTS seems to have the most preferred overall advantages:

[Set 1]

Which amounts to meaning (5/32). [11-d] Sq rt of 25 /32

Which amounts to sublimism (15/17?). [10-d] $15 \wedge 1 / 17$

Which amounts to meaningful systems (25/32?). [9-d] $25\wedge1 / 32$

Which amounts to philosophical technology (225/17?). [8-d] $15\wedge2 / 17$

Which amounts to preferences and dimensional worlds (625/32?). [7-d] $25\wedge2 / 32$

LINEAR / CYCLICAL

THE WHEEL OF THE WORLD

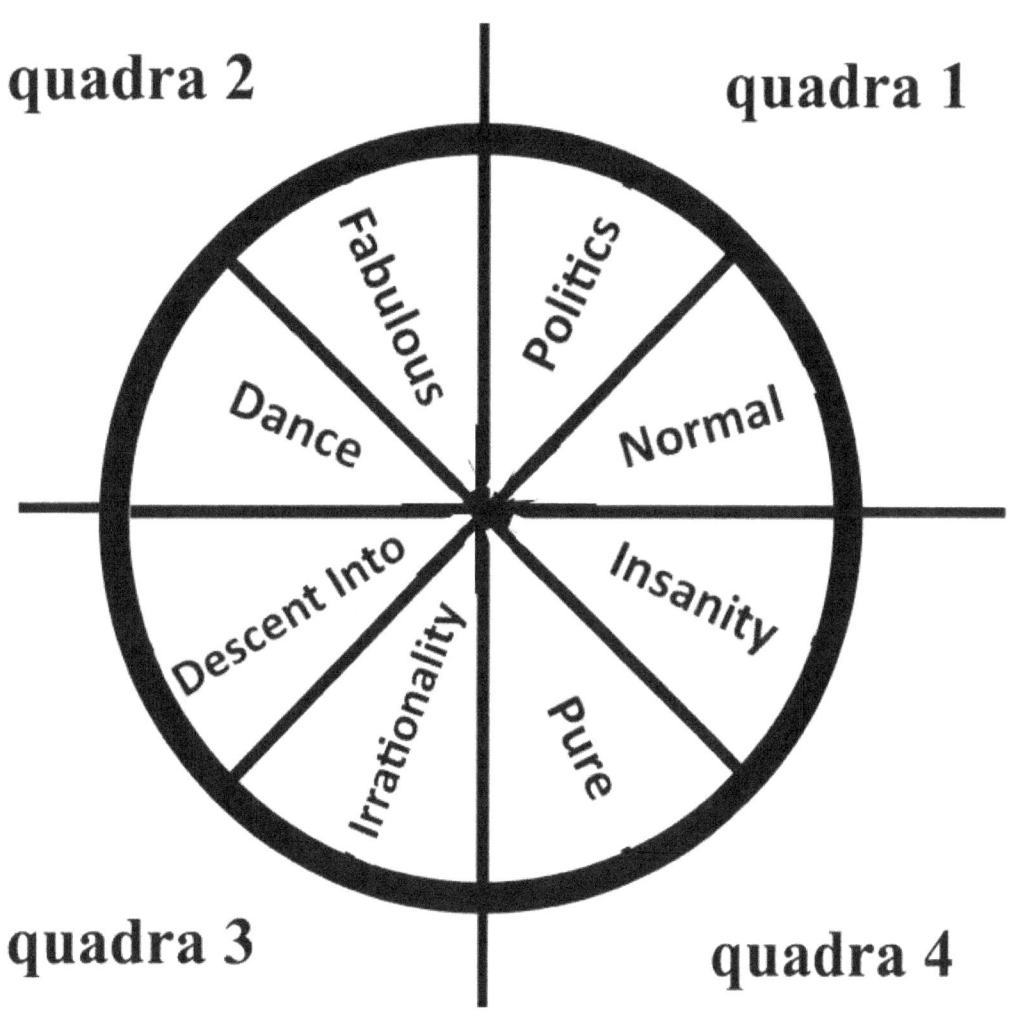

COHERENT MODELS

SUMMARY MODELS

Dual-Descriptive Model: Ideas or abstraction might be 4-d, Goals or limits might be 5-d, To simplify, unique things or paradoxes might be seen as 6-d, infinite singularities or descriptions of one thing might be 7-d, loops or perfected physical systems might be 8-d, the TOE or theorizing might be 9-d, a collection of alternate systems of sufficient perfection or a wish to be immortal might be 10-d, and impossibility or virtual reality might be 11-d.

Energy Model: The universe is 9/25 mathematically 0-energy, 5/25 negative energy, and 11/25 infinite energy.

Grounded Theory Model: Anything except kites and spacecraft someone might say.

Negative Dimensional Model: The universe requires five negative dimensions, which are not as obviously useful as the highest dimensions, though they are fundamental for language.

Objectivity Model: Objectivity can be 10-dimensional. For example, if I want to be immortal, wishing for immortality might be 10-dimensional. However, abstraction is more likely to be negative dimensional or perhaps up to 4-dimensions.

COHERENT MODELS / *"SUPERINTELLIGENCES"*

T.O.E.

- KNOWLEDGE: Results (1,2,3...) = Eff + Difference
- PERPETUAL MOTION: Results = Eff (1,2,3...) + Difference
- FUNCTION SPECTRUM: Results = Eff + Difference (1,2,3...)
- Unified—!

Anti-Theory:

- Anti-Thing (1,2,3...) <= Difference - Efficiency
- Anti-Thing <= Difference (1,2,3...) - Efficiency
- Anti-Thing <= Difference - Efficiency (1,2,3...)

Efficiency:

- Efficiency (1,2,3...) >= Results – Difference
- Efficiency >= Results (1,2,3...) – Difference
- Efficiency >= Results – Difference (1,2,3...)

Anti-Efficiency:

- Anti-Efficiency (1,2,3...) <= Difference - Results
- Anti-Efficiency <= Difference (1,2,3...) - Results
- Anti-Efficiency <= Difference - Results (1,2,3...)

Difference:

- Difference (1,2,3...) >= Results – Efficiency
- Difference >= Results (1,2,3...) – Efficiency
- Difference >= Results – Efficiency (1,2,3...)

Anti-Difference:

- Anti-Difference (1,2,3...) <= Efficiency - Results
- Anti-Difference <= Efficiency (1,2,3...) - Results
- Anti-Difference <= Efficiency - Results (1,2,3...)

Forces:

- # Forces (1,2,3...) = # Dimensions - # Antiforces
- # Forces = # Dimensions (1,2,3...) - # Antiforces
- # Forces = # Dimensions - # Antiforces (1,2,3...)

Antiforces:

- # Antiforces (1,2,3...) = # Dimensions - # Forces
- # Antiforces = # Dimensions (1,2,3...) - # Forces
- # Antiforces = # Dimensions - # Forces (1,2,3...)

Dimensions:

- # Dimensions (1,2,3...) = # Forces + # Antiforces
- # Dimensions = # Forces (1,2,3...) + # Antiforces
- # Dimensions = # Forces + # Antiforces (1,2,3...)

Anti-Dimensions:

- # Anti-Dimensions (1,2,3...) = # Antiforces - # Forces
- # Anti-Dimensions = # Antiforces - # Forces (1,2,3...) +
- # Anti-Dimensions = # Antiforces (1,2,3...) - # Forces

Disintegral (same as before, by De Morgan's Rule):

- WAR EQUATION: Disintegral (1,2,3...) = Efficiency – Difference
- EFFICIENCY SPECTRUM: Disintegral = Efficiency (1,2,3...) – Difference
- GENERAL AND SPECIAL TRANSLATION: Disintegral = Efficiency – Difference (1,2,3...)

Anti-Disintegral or Abstract Efficiency (Same as Antitheory, modifying the change after De Morgan's):

- Anti-Disintegral (1,2,3...) = Difference – Efficiency
- Anti-Disintegral = Difference (1,2,3...) - Efficiency
- Anti-Disintegral = Difference - Efficiency (1,2,3...)

Super-Disintegral:

- Super-Disintegral (1,2,3...) = Inf Efficiency – Inf Difference
- Super-Disintegral = Inf Efficiency (1,2,3...) – Inf Difference
- Super-Disintegral = Inf Efficiency – Inf Difference (1,2,3...)

Anti-Super-Disintegral:

- Anti-Super-Disintegral (1,2,3...) = Inf Difference – Inf Efficiency
- Anti-Super-Disintegral = Inf Difference (1,2,3...) - Inf Efficiency
- Anti-Super-Disintegral = Inf Difference - Inf Efficiency (1,2,3...)

Min Results:

- Min Results (1,2,3...) = (Max Eff / 2) + Diff
- Min Results = (Max Eff (1,2,3...) / 2) + Diff
- Min Results = (Max Eff / 2) + Diff (1,2,3...)

Max Results:

- Max Results (1,2,3...) = Min Eff + Diff
- Max Results = Min Eff (1,2,3...) + Diff
- Max Results = Min Eff + Diff (1,2,3...)

Min Efficiency:

- Min Eff (1,2,3...) = Results - Diff
- Min Eff = Results (1,2,3...) - Diff
- Min Eff = Results - Diff (1,2,3...)

Max Efficiency:

- Max Eff (1,2,3...) = (Min Results - Diff) X 2
- Max Eff = (Min Results (1,2,3...) - Diff) X 2
- Max Eff = (Min Results - Diff (1,2,3...)) X 2

Flying Max Results

- Flying Max Results (1,2,3...) = (Min Eff) + 2 Eff - 1
- Flying Max Results = (Min Eff (1,2,3...)) + 2 Eff - 1
- Flying Max Results = (Min Eff) + 2 Eff (1,2,3...) - 1

Flying Min Results

- Flying Min Results (1,2,3...) = (Max Eff / 2) + 2 Eff - 1
- Flying Min Results = (Max Eff (1,2,3...) / 2) + 2 Eff - 1
- Flying Min Results = (Max Eff / 2) + 2 Eff (1,2,3...) - 1

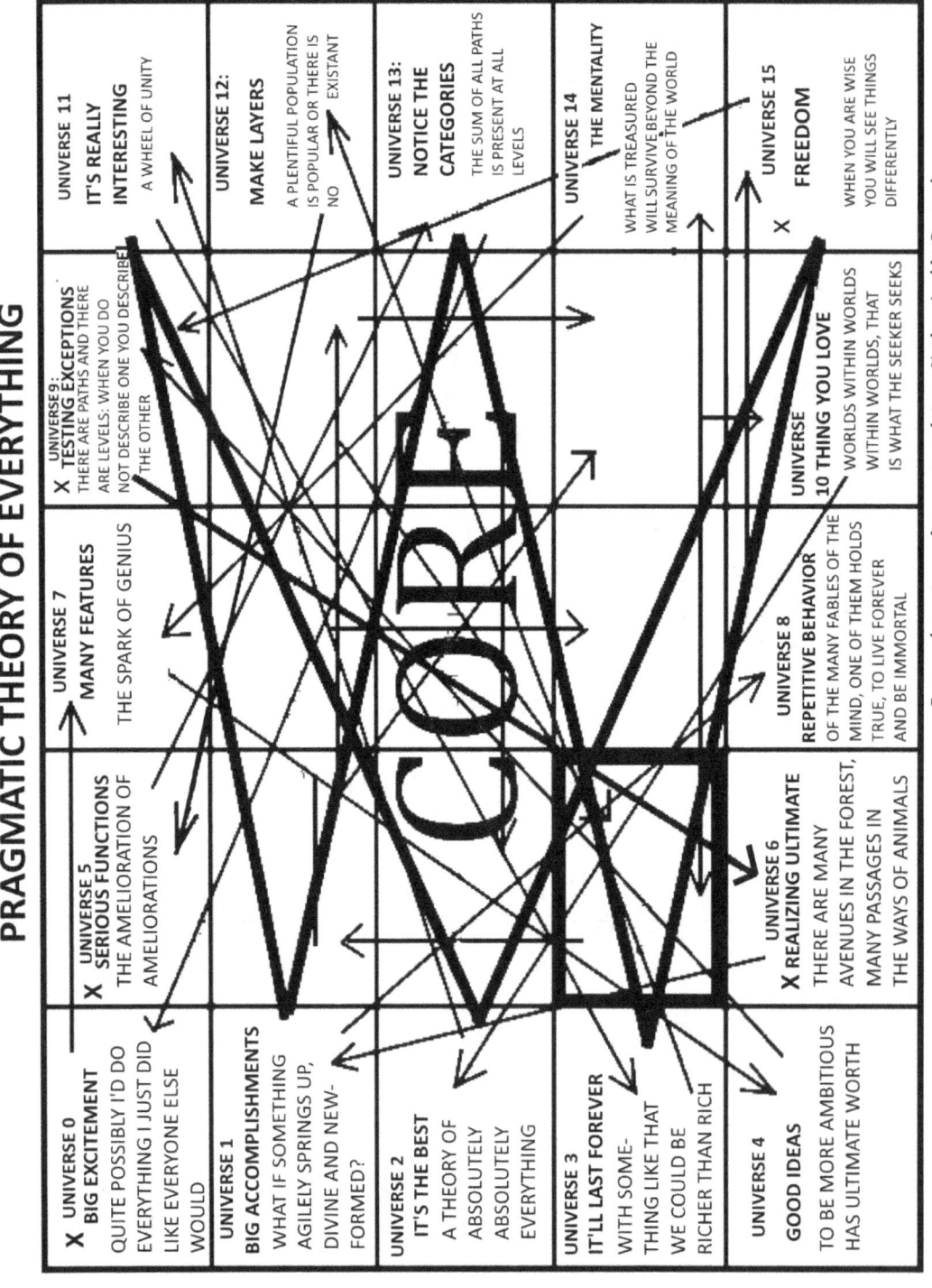

SUPERHUMANISM

META-NIRVANA

Perpetual motion is not deathly (nature)... Medicine is not continuous (reliable)... Ideas are not perpetual (living)... Nirvana is not medicine (helpful)... Infinite variations is not an idea (timeless)... Souls are not nirvana (clarity)

THE FORGE OF CONSTANTS

CONCEPTUAL DIMENSIONS 8

TOTAL CATEGORIES 2097152 (FYI this appears to be a 2 base 16)

PHYSICAL DIMENSIONS 2097019.415

NUMBER OF IDEAS 2097145.831

ALTERNATE IDEAS -2 524287.4578 IMMORTALS

ALTERNATE IDEAS -1 1048573.916 ARCHETYPE

ALTERNATE IDEAS +1 4194293.662 BASE IDEA

ALTERNATE IDEAS +2 8388587.325 INSULTING

SPECIES: 2097026.583

SPECIES - 2: 524257.3959

SPECIES - 1: 1048513.792

SPECIES + 1: 4194053.167

SPECIES + 2: 8388106.334

At Physical Dimensions 2097152.169 the number of ideas becomes exactly 2097146

Separately modifying CONCEPTUAL dimensions to 2097152 results in exactly 2097151 IDEAS when 2097152 TC is inputted. Unfortunately some of the other results were numbed out.

FORTUNATEMEN CONCEPT

A PRIMARY ORDER OF FACTS ON THE SUBJECT

Within the 3rd dimension, Divine Art (3-d metaphysical creation) and Diabolical Science (e.g. 3-d yogis) were the previous alternatives to Longevity Evolution (ideal survival).

Evolution subtly implies the failure of God and demons, and either the success of ideal survival, or the incrementation into the 4th dimension where virtual reality magic (real information), time-travel and immortality (2-d time), teleportation (wormholes), and invisibility (manifesting in higher dimensions) are likely powers.

In the dimensional continuum / metaphysico-physical-continuum / Flatland theory the number of dimensions defines the physical laws for the whole universe. If there is another universe, it would have different laws about the fundamentals of dimensions.

Recursive stages: 1. Evolution, 2. Consciousness, 3. Meaning, 4. Perfection.

Sometimes a world is born just because its more efficient, and sometimes it is born because it fulfills a fantasy. What creates multiple worlds must have a good idea, and a reason for their separation or association.

1.5 --- points with a translation into 2-d correspondence (Graph Theory).

2.5 --- holograms and shadows and noise.

3.5 --- time-travel, immortality, invisibility, and teleportation.

Imagining further increments is a matter of objectifying the individual properties which constitute the difference between the dimensional levels. Conveniently, I have provided a typology in which those three increments each have one category of relation per following dimension.

In the 1.5 - 2nd dimension, we only measure one value (graphing), and the realization of the degrees between 1.5 and 2nd becomes something like triangulation preceding 2-d and 3-d figures. In the 2nd dimension two-d and three-d figures are roughly the same. In the 2 - 2.5th dimension we take complexity and perfection (the typology of the 2-d), and reach the perception of holograms, shadows, and noise. In the 2.5 - 3rd dimension, we realize the full potential of holograms, shadows, and noise, leading up to physical objects, organics, liquids, and time-perception. In the 3rd to the 3.5 we find a relation between physical objects, organics, liquids, and time-perception, and perception of the 4-d meaning time-travel, immortality, invisibility, and teleportation.

(PROOFS OF SOME DIMENSIONS)

Proof in the first dimension: Something exists.

Proof in the second dimension: Two seperate things exist.

Proof in the third dimension: Sculptures exist.

Proof in the fourth dimension: Time-travel exists.

Proof in the 5th dimension: Choices-about-time-travel-while-time-travelling exist.

Proof in the 6th dimension: Judgements-about-how-to-time-travel-negotiated-between-two-seperate-time-travellers-as-they-travel-differently exist.

Proof in the 7th dimension: Timeless judgement between immortals. In other words, an objective timeless realm.

Proof in the 8th dimension: The creation of a new timeless realm within an old timeless realm. Pure thought / being.

Proof in the 9th dimension: New tools for using the timeless realm. Interface / Archetypes.

Proof in the 10th dimension: Paradoxical tools. "Material complexity." Panopticon.

Proof in the 11th dimension: Super-sentience, milk of life.

Proof in the 12th dimension: Finite creation, ascendant thought.

Proof in the 13th dimension: Super-identity, trans-realism.

Proof in the 14th dimension: Far-seeing, obliteration.

Proof in the 15th dimension: Dimensional mystery, knots of time and space.

Proof in the 16th dimension: Highest precept, the constructiv.

Proof in the 17th dimension: The culture of space, the meaningful art.

Proof in the 18th dimension: Pure mathematics.

Proof in the 19th dimension: Holes in time and space.

Proof in the 20th dimension: Order.

Proof in the 21st dimension: Mechanic.

See also: Philosophy's Bulletin-Points for Analytical Critique

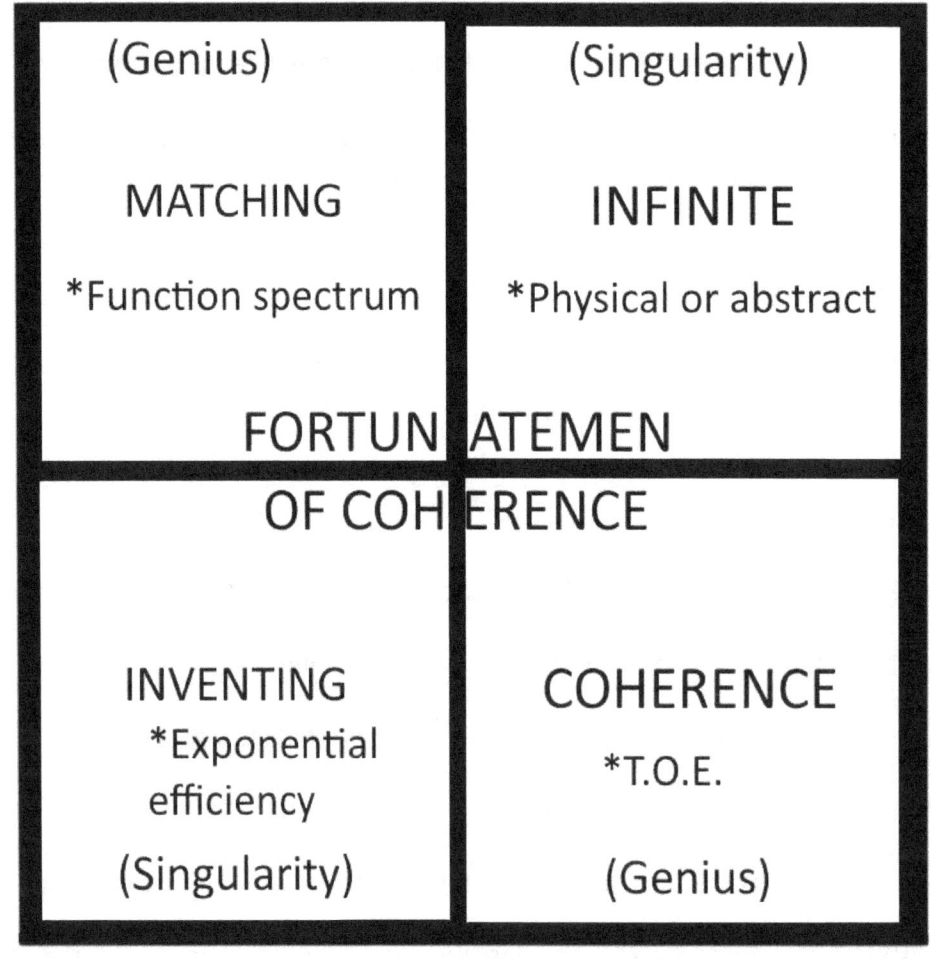

MAGIC OF THE FORTUNATEMEN

A HISTORICAL MODEL OF DEVELOPMENTAL COHERENCE

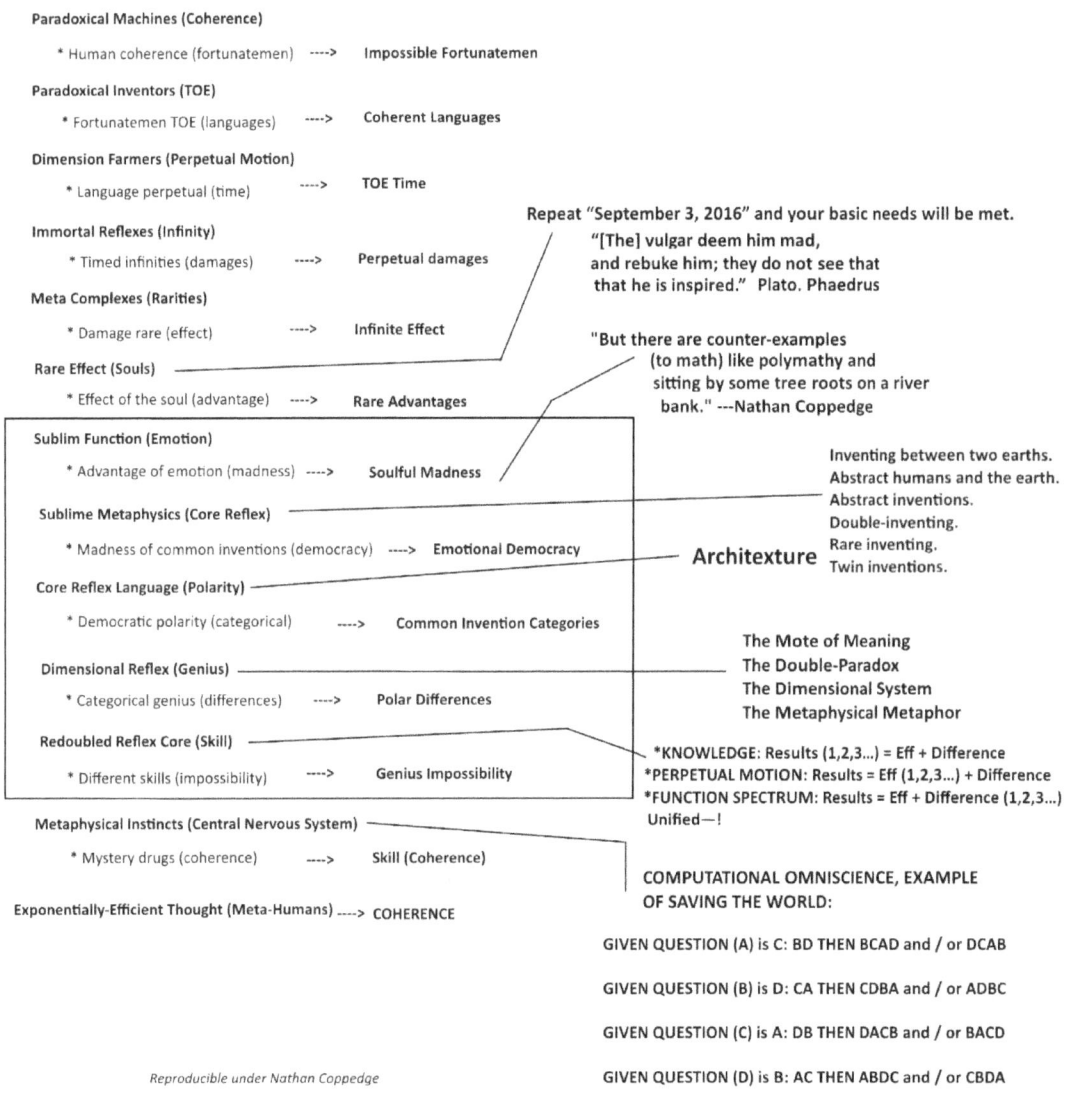

Reproducible under Nathan Coppedge

Consider: 0.2640625

INTELLIGENCE RELATIVE TO THE NUMBER OF PHYSICAL DIMENSIONS & TOE CATEGORIES
"ALIEN I.Q." PROPORTIONS

D	TCat	Intelligence		D	TCat	Intelligence	
2	1	negligible		4	1	negligible	
2	9	1.2	PERFECT 4 IDEAS	4	81	1.013	PERFECT 40 IDEAS
2	25	1.8181		4	625	1	
2	49	2.47	PERFECT 22 IDEAS	4	2401	1.035	PERFECT 1198 IDEAS
2	81	3.13		4	6561	1.1	
2	121	3.793	PERFECT 56 IDEAS	4	14641	1.2	PERFECT 7316 IDEAS
2	169	4.457		4	28561	1.3	
2	225	5.12	PERFECT 106 IDEAS	4	50625	1.41	PERFECT 25306 IDEAS
2	289	5.787		4	83521	1.52	
2	361	6.453	PERFECT 172 IDEAS	4	130321	1.635	PERFECT 65152 IDEAS
2	441	7.119		4	194481	1.753	
2	529	7.785	PERFECT 254 IDEAS	4	279841	1.872	PERFECT 139910 IDEAS
2	625	8.45		4	390625	1.99	

D	TCat	Intelligence	
3	1	negligible	
3	27	1	
3	125	1.07	
3	343	1.235	
3	729	1.429	
3	1331	1.634	
3	2197	1.84	NO IMMEDIATELY PERFECT IDEAS
3	3375	2.059	
3	4913	2.275	
3	6859	2.493	
3	9261	2.71	
3	12167	2.93	
3	15625	3.15	

D = CONCEPTUAL DIMENSIONS TC = MATH CATEGORIES

INTELLIGENCE = IDEAS / PHYSD
This gives intelligence relative to the number of physical dimensions.

PHYSICAL DIMESIONS =
Number of Math Categories Minus Negative Dimensions then Minus The Neutral (Inner Area) = (Tcategories - Droot of T) - [(Droot of T - D) ^D]

IDEAS = Number of Math Categories Minus Negative Dimensions = (Tcategories - Nroot of T)

REPRODUCIBLE UNDER NATHAN LARKIN COPPEDGE

...

Nathan Coppedge (b. 1982) Philosopher, Artist, Inventor, Poet and member of the international honor society for philosophers, is best known for his writings on philosophy and for his perpetual motion designs & theory, and abstract art called Hyper-Cubism. He lives near Yale University.

www.ingramcontent.com/pod-product-compliance
Lightning Source LLC
Chambersburg PA
CBHW080446220526
45465CB00007B/2782